Adel Ali Mohamed

Marble/ Granite composite

Adel Ali Mohamed

Marble/ Granite composite

to remove Lead and Cadmium from contaminated water

Noor Publishing

Imprint

Cover image: www.ingimage.com

Publisher:
Noor Publishing
is a trademark of
Dodo Books Indian Ocean Ltd. and OmniScriptum S.R.L publishing group

120 High Road, East Finchley, London, N2 9ED, United Kingdom
Str. Armeneasca 28/1, office 1, Chisinau MD-2012, Republic of Moldova, Europe
Managing Directors: Ieva Konstantinova, Victoria Ursu
info@omniscriptum.com

Printed at: see last page
ISBN: 978-620-9-05337-5

1- Introduction

Water is a vital natural resource that we use daily for development and drinking. Water that is safe to drink is essential for everyone's health, everywhere. Water is a ubiquitous solvent and a primary cause of infectious illnesses. The primary sources of water contamination include runoff from residential and commercial wastewater, air deposition, storage tank leaks, ocean dumping, and radioactive waste. Industrial trash and heavy metals that have been discarded end up in lakes and rivers, where they can harm both people and wildlife Industrial wastes contain toxins that can lead to acute poisoning, immunosuppression, and reproductive failure. Contaminated water can spread infectious diseases such as cholera, typhoid fever, gastroenteritis, diarrhea, vomiting, skin infections, and kidney disease **(Pal et al., 2018).** While certain metals naturally oppose the vital functions of other elements, human activity has resulted in a substantial pollution load that surpasses the aquatic environment's ability to clean itself. In small amounts, the body requires a variety of trace elements and heavy metals like zinc, copper, iron, manganese, and cobalt **(Lane et al., 2000)**; while they are toxic at higher concentrations **(Chronopoulos et al., 1997).** However, an additional 12 harmful heavy metals, including lead, mercury, cadmium, chromium, and nickel, are detrimental to health. The poisonous heavy metal ions first enter the human body through the food chain **(Merzouk et al., 2011)**. Because heavy metals are not biodegradable, they stay in the human body permanently for a considerable amount of time **(Hahladakis et al., 2013; Diagomanolin et al., 2004).** Two prevalent hazardous heavy metals to which man is exposed more and more are lead and cadmium, because of their wide range of applications in materials and technologies that are essential to human survival. The highest permissible content of Pb^{2+} in drinking water was limited to 20 mg/ml by the World Health Organization's worldwide water quality regulations **(World Health Organization, 2000).** Cadmium is one of the more poisonous metals, with intake causing several symptoms such as high blood pressure, renal damage, and red blood cell destruction **(Ghazy, 1995; Ghazy et al., 2008b)**. The methods for treating wastewater include separation, coagulation/flocculation, precipitation, ion exchange, electrocoagulation, reverse osmosis, and adsorption. Adsorption has been confirmed to be excellent for different methods of water purification in terms of ease of application, cost, simplicity of design, and viability for in situ treatment of underground and surface water. Adsorption is a surface phenomenon where molecules, atoms, or ions from a gas, liquid, or dissolved solid cling to the adsorbent's active site to form an adsorbate film on the adsorbent's surface. The process of removing a substance from a liquid or gas by means of a solid phase that results in a larger concentration (or accumulation) of removed adsorbate molecules on the adsorbent surface compared to that in the bulk of the solution is known as adsorption, as shown in Figure. 1 **(Soliman and Moustafa, 2020).** Furthermore, because adsorption does not require expert maintenance or equipment, it can be used in rural regions **(Mizuta et al., 2004; Batheja et al., 2009)**. General Requirements of Adsorbents High abrasion resistance, high thermal stability, and tiny pore sizes are necessary for adsorbents because they increase exposed surface area and, thus, adsorption surface capacity. Additionally, the adsorbents need to have a unique pore structure that allows the gaseous vapors to be transported quickly **(Dureja et al., 2007).** Some examples are used as adsorbents, such as clay **(Zacaroni et al.,2015)** , activated carbon**(Kobya et al.,2005)** alumina,chitosan zeolites (**Nibou et al., 2010**) and peat moss .Granite powder and marble. Granite powder, a granulated natural rock (NR), which is volcanic granite rock generated from magma and is commercially available, was gathered and utilized as the adsorbent material. Although granitic magma has a variety of possible sources, marble is an inorganic adsorbent that is frequently found in Egypt, China, India, Italy, Spain, and Turkey is marble powder. It has iron oxide, silica, calcium carbonate, clay, potassium, and sodium salts. It is an economical inorganic sorbent that can be used to remove water contaminated with dyes and heavy metals **(Diouri et al., 2014).** Marble waste is used for sorptive flotation to remove lead, copper, and cadmium ions from an aqueous solution **(Alwared et al., 2019).** Powdered marble waste (PMW) is used to remove Zn(II) ions from aqueous solutions and water samples(**Ghazy et al., 2008**) , and the use of marble in pebble size to remove excessive iron from groundwater **(Xin, 2019). Alwared et al.(2019)** removed 99.95% of Pb (II), 84.58% of Cu (II), and 78.70% of Cd(II) ions using dry marble powder by. flotation in a comparatively short amount of time. **Tozsin(2016)** extracted more than 80% of the Cd, Cr, Cu, Ni, Pb, and Zn from the acid drainage of mines produced by Cu (II) flotation tailings by using waste from marble cutting. **Wazwaz et al.(2019)** investigate the use of adsorption to remove heavy metals (Cu, Zn, Mn, and Cr) from polluted water. As an adsorbent, three types of soil (silty, sandy, and clay) and three types of marble powder (pure, impure, and marble-granite mix) were utilized. Granite was used to study the adsorption behavior of copper (II) ions from an aqueous solution as a function of several parameters, including dosage, initial concentration of copper (II) ions, contact time, and temperature **(Hussain et al., 2012).** Granite, beach sand, and river sand are all characterized as potential adsorbents for the treatment of water or wastewater **(Familusi et al., 2021).**

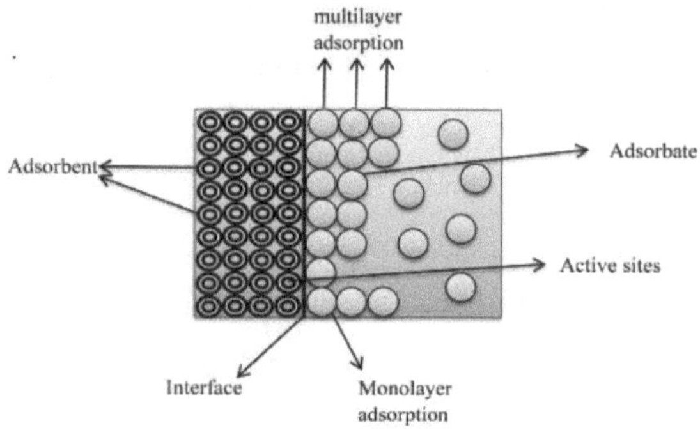

Figure (1): Schematic diagram of the adsorption process.

2.1. Reagents and chemicals

Deionized water was used throughout the study to carry out all the experiments. The preparation of standard solution stocks containing 1000 ppm of Pb^{2+} and Cd^{2+} ions involved by dissolving cadmium stock solution [Cd $(NO_3)_2$ in HNO_3 0.5 mol/l, the concentration of Cd^{2+} = 1000 ppm] and lead stock solution [Pb $(NO_3)_2$ in HNO_3 0.5 mol/l, the concentration of Pb^{2+} = 1000 ppm]. To stop hydrolysis, the stock solutions were made more acidic. The working metal ion solution was prepared from the stock solution by dilution. A wastewater sample was used to evaluate the potential efficiency of the developed adsorbents to eliminate heavy metals from wastewater. A water sampler with a handle of 500 ml was used for collecting wastewater from the lab disposal of the Kima factory. Samples were clarified by filter paper (Whatman-42), collected in one liter polyethylene bottles, and preserved in the refrigerator at 5 °C until the experiment.

2.2. Characterization of granite adsorbent, marble adsorbent, and mixture of granite: marble (2:1)

The adsorbent was examined using SEM-EDX (scanning electron microscopy-energy dispersive X-ray) (SEM-EDX, JSM-IT200) XRD (using a PAN analytical X-ray Diffraction apparatus model X'Pert PRO with Secondary Monochromatic Cu-radiation), and FTIR (the Shimadzu Tracer 100 (made in Japan)). The N_2 adsorption capacity was used to compute the specific surface area (SSA) using the Brunauer-Emmett-Teller (BET) equation to calculate the porosity and surface area of granite, marble, and granite marble mixture adsorbent samples.

2.3. Collection of the Samples

2.3.1. Granite and marble samples

Granite and marble ores were collected from Egypt (Aswan City). A laboratory crusher was used to crush and powder the materials, and after that, a sieve analysis was performed using a 45-micron sieve. The chemical composition of the granite and marble samples was identified using XRF, as indicated in Table (1).

Table (1): Chemical composition of granite and marble samples determined by XRF.

Composition	granite	marble
	(ppm)	(ppm)
Al_2O_3	91500	25
CaO	6500	551600
Fe_2O_3	11000	25
K_2O	46550	30
MgO	500	40
MnO_2	2450	30
Na_2O	13500	50

P_2O_5	550	30
SiO_2	551	1500
TiO_2	1800	20
CO_2	----	431200
H_2O	------	40

2.3.2. Real wastewater sample

The wastewater sample was collected from the lab disposal of Kima Factory (Kima 2). Samples were purified by filter paper (Whatman 42).

2.4. Preparation of adsorbent samples

Taking the size under 45 microns, the following samples were prepared (wt. by wt.). Five samples were prepared: Sample No. 1 granite, Sample No. 2 marble, Sample No. 3 granite: marble (1:1), Sample No. 4 granite: marble (1:2), and Sample No. 5 granite: marble (2:1).

2.4.1. Adsorption of Lead and Cadmium

Each sample weighed 0.5 g was agitated at 250 rpm using 50 ml of a standard solution (50 ppm Pb^{2+} and Cd^{2+}) for 1 hour at a stirring rate of 250 rpm, the pH of the mixture was adjusted to pH7 using NaOH or HNO_3 solution at temperatures of 25, 50, and 60 °C. After filtering the mixture, an atomic absorption spectrophotometer was used to measure the metal ion concentration in the filtrate. Every adsorption experiment has been carried out three times, and the standard mean was determined. The percentage of metal ions removal adsorbed was calculated by the following equation:

$$\% R = [(C_o - C_e) / C_o] * 100 \qquad (1)$$

Where: C_o = initial concentration of adsorbate (ppm)

C_e = the final concentration (ppm)

2.5. Batch adsorption experiments on a mixture of sample No. 5 granite: marble (2:1)

Experiments using a batch technique were done to remove Pb^{2+} and Cd^{2+} ions using granite: marble (2:1) adsorbent. The effects of pH, initial adsorbate concentration, adsorbent dosage, temperature, and contact time were investigated in batch adsorption experiments. An atomic absorption spectrophotometer was used to measure the filtrate's metal ions. Every adsorption experiment has been carried out three times, and the standard mean was determined.

2.5.1. Initial metal concentration effect

A mass of 0.1 g granite: marble (2:1) was stirred (250 rpm) using 50 ml of a standard 50 ppm heavy metal ions Pb^{2+} and Cd^{2+} for 2 hours at pH 6, at different metal concentrations (10, 25, 50, 75, and 100 ppm) as the initial concentration at constant temperature (25°C). The mixture was extracted and filtered. The concentration of metal ions in the filtrate was measured using an atomic absorption spectrophotometer.

2.5.2. pH effect

The following adsorption studies were carried out to ascertain the impact of solution pH: 0.1 grams of granite: marble (2:1) was stirred (250 rpm) using 50 ml of a single-ion standard solution that 50 ppm Pb^{2+} and Cd^{2+} for 2 h. The pH was 2, 5, 6, 7, and 9 (pH adjusted using HNO_3 (HCl) or NaOH solution).

2.5.3. Adsorbent dosage effect

Varying masses (0.025,.05,0.1,0.2,0.25.0.5, and 1.0 g) of granite: marble (2:1) were mixed for two hour using 50 ml of standard metal solutions containing 50 ppm of single ions (stirring at 250 rpm) and adjusted at pH 6 for Pb^{2+} and Cd^{2+}. After filtering the mixture, the amount of metal ions present in the filtrate was determined.

2.5.4. Time contact effect on heavy metal adsorption

0.1 g of granite: marble (2:1) was stirred using 50 ml of a standard solution containing 50 ppm of Pb^{2+} and Cd^{2+} single heavy metal ions at pH 6. The initial concentration of the metal was 50 ppm. Temperature did not change from 25 °C for contact times of 10, 30, 60, 90, and 120 min.

2.5.5. Temperature effect on heavy metal adsorption

0.1 g of granite: marble (2:1) was stirred using 50 ml of a standard solution containing 50 ppm of Pb^{2+} and Cd^{2+} single heavy metal ions at pH 6. At temperatures, the influence of the adsorption temperature was studied at 25, 40, 50, and 60 °C.

3. Results and Discussion

3.1. Characterization of granite, marble, and mixtures

3.1.1. XRD Analysis

Figure 2 depicts an XRD analysis to determine the structural features of marble, granite, and mixtures. XRD of granite confirmed the presence of quartz (SiO_2), biotite ($K(FeMg)_3AlSi_3O_{10}(OH)_2$), almandine ($Fe_3Al_2$ $(SiO_4)_3$, albite ($NaAlSi_3O_8$) and calcite ($CaCO_3$). High intensity of quartz (SiO_2) at 2θ = (23.9693, 50.0721 and 68.1041°), biotite ($K(FeMg)_3AlSi_3O_{10}(OH)_2$), peak at 2θ = (8.7434°), almandine (Fe_3Al_2 $(SiO_4)_3$ peak at 2θ= (26.5640 and 27.9280°), albite ($NaAlSi_3O_8$) peak at 2θ= (13.6574, 22.0358, 23.5382 and 25.5084°). The Mineral Name(JCPDS Card N), Quartz SiO_2 (83-0539), Albite Na Al Si_3O_8(10-0393), Alamandine $Fe_3Al2(SiO_4)_3$(85-2495) and Calcite $CaCo_3$(87-1863)(**Yarrakula ,2019)** .The same minerals have also been found in other studies on the geochemical characterization of different kinds of granite mining soils **(Gao et al., 2017)** . The highest peak appeared at 2θ = 29.3660°, indicating that the diffraction angle is calcite ($CaCO_3$). Calcite and dolomite peak attributes are consistent with those reported by **(Gunasekaran et al., 2007).**

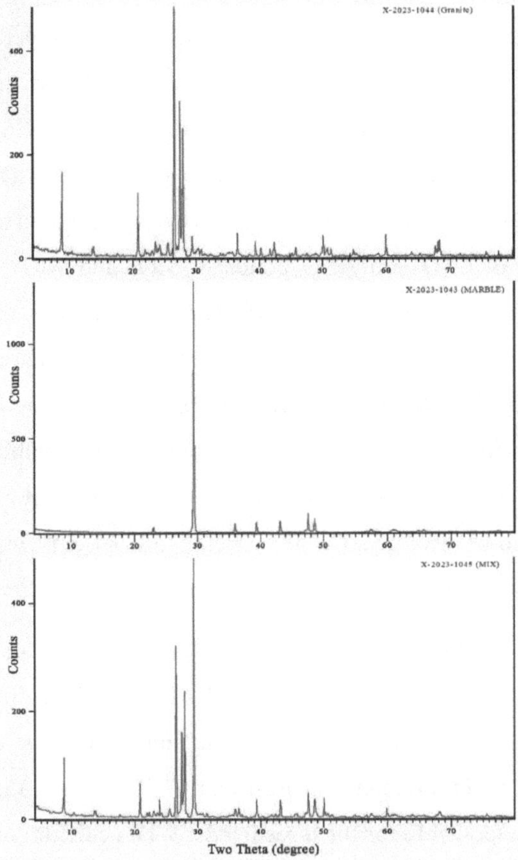

Figure (2): XRD of granite, marble and granite: marble (2:1).

3.1.2. FTIR analysis

The analytical technique of FTIR spectroscopy is helpful for determining functional groups and

describing the details of covalent bonds. The IR spectra of granite, marble and granite and marble (2:1) were illustrated in (Figure: 3). Granite sample showed three peaks at 436.38 cm-1, 586.61 cm^{-1}, and 776.32 cm^{-1}, assigned to ring vibrations of silicates **(Sitarz et al., 2000)**, while the presence of a band at 916 cm^{-1}, assigned to asymmetric stretching of Si-O and Al-O bonds in SiO_4 and AlO_4 tetrahedrons, respectively **(Handke et al., 1993).** For marble the stretching vibrations of carbonate molecules C=O stretching at positions 1799.88, 875.55, and 710.92 cm-1 are responsible for the three bands. When calcium hydroxide and atmospheric carbon dioxide combine, carbonate molecules are created **(Handke et al., 1993; Yılmaz et al., 2008)**. For mixture shows three peaks at 436.38 cm^{-1}, 586.61 cm^{-1}, and 776.32 cm^{-1}, assigned to ring

vibrations of silicates, and the presence of a band at 916 cm^{-1}, assigned to asymmetric stretching of Si-O and Al-O bonds in SiO_4 and AlO_4 tetrahedrons, respectively. The stretching vibrations of carbonate molecules C=O stretching at positions 1799.88, 875.55, and 710.92 $cm-^{1}$ are responsible for the three bands. When calcium hydroxide and atmospheric carbon dioxide combine, carbonate molecules are created.

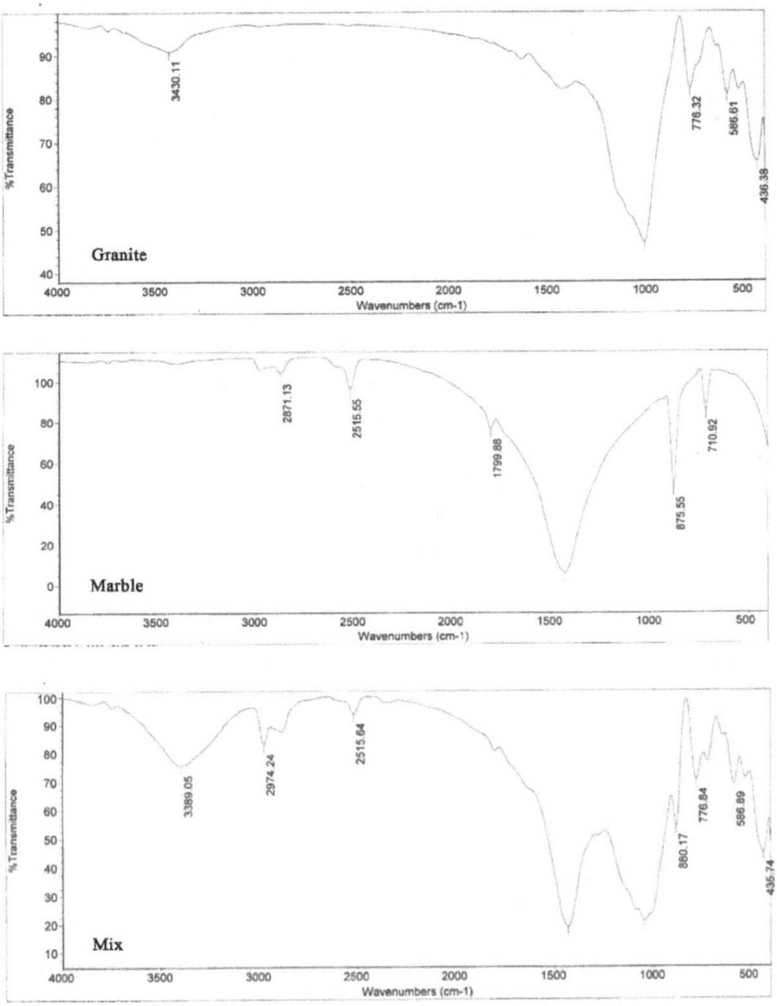

Figure (3): FTIR of granite, marble and granite: marble (2:1).

3.1.3. Analysis of Surface area (BET, and BJH)

In this study, the surface area was determined using the Brunauer Emmet-Teller (BET) method **(Brunauer et al., 1938)**. The method is based on the physical adsorption of gases on the external and accessible internal surfaces of a porous material.

Table (2): The surface area and pore size of granite, marble, and granite: marble (2:1).

	granite	marble	granite: marble (2:1)	Unite
Surface area as (BET)	2.3177	1.0437	2.0435	$m^2 g^{-1}$
Total pore volume	0.0165	0.0132	0.0178	$cm^3 g^{-1}$
Average pore diameter	28.526	50.625	34.972	Nm

The results granite: marble (2:1) for Surface area as, (BET) and average pore diameter were the average values of granite and marble, but for average pore diameter give a value higher than the average of the two separately.

3.1.4. SEM, EDX examinations

The micro-morphology and chemical composition of granite, marble, and mixture adsorbents were examined by scanning electron microscopy and energy-dispersive x-ray spectrometry. The morphological features of particle microstructure were evaluated by SEM. Figures (4) show the SEM picture for granite, marble, and a mixture of granite: and marble. Granite showed a heterogeneous microstructure where particles are agglomerated and dense with different sizes and irregular shapes. The samples have heterogeneous textures with many cracks. The heterogeneity of the microstructure could be related to the presence of several crystalline mineral phases, particle size distribution, and shape of granite waste particles. It was proven that the geopolymerization occurs at the surface of the aluminosilicate particles **(Favier et al., 2013).** Marble which illustrates the presence of well-delineated particles with irregular and angular shapes, and with a wide range of size. The images display the uniformity of the granite and marble particles, along with a few interstitial spaces. The images display the uniformity of the granite and marble particles, along with a few interstitial spaces. Figures (5) show the EDX spectrum of granite, marble, and a mixture of granite: marble (2:1). From the EDX spectrum analysis of granite, it is found that the major elements present in granite are silicon, aluminum and iron, and small traces of metals and metalloids. The major constituent of marble is calcium. The percentages of two constituents (silicon and calcium) are obvious in the mixture EDX. The major element of marble is calcium while silicon is the main element of granite. The two composite materials have great differences in structure and composition which explain that they have great difference in manufacturing parameters and electrical and mechanical properties. Table (2) shows an analysis of granite, marble, and granite, marble (2:1) by EDX spectrum.

Table 3: Chemical composition of granite, marble, and granite: marble (2:1) samples determined by EDX

Elements %	granite	marble	Granite: marble (2:1)
C	9.83	13.43	9.39
O	46.9	53.33	49.98
Na	1.33	0.03	2.46
Al	0.5	0.26	0.23
Mg	5.45	0.13	5.92
Si	22.85	0.06	18.59
K	4.14	----	1.65
Ca	1.94	31.89	10.04
Ti	0.6	0.04	----
Fe	5.03	----	1.75
Cu	1.42	0.84	-----

granite	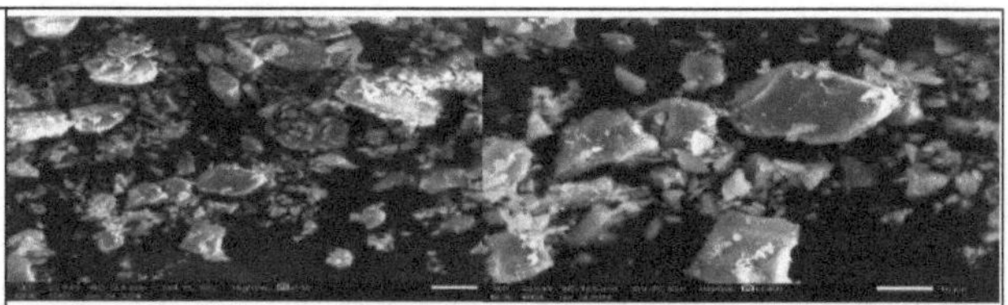
marble	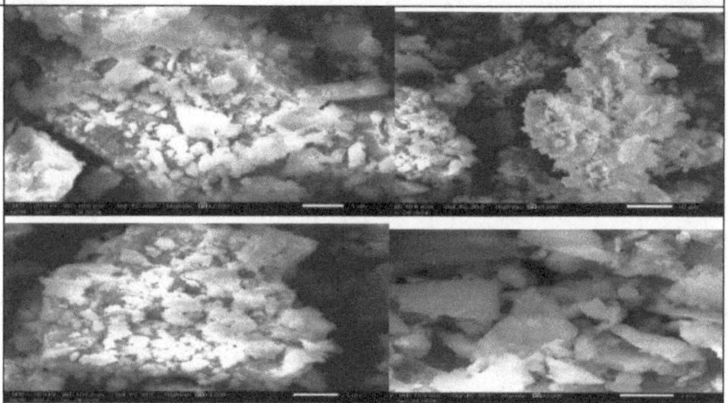
granite: marble (2:1)	

Figure (4): SEM picture of granite, marble, and granite: marble (2:1)

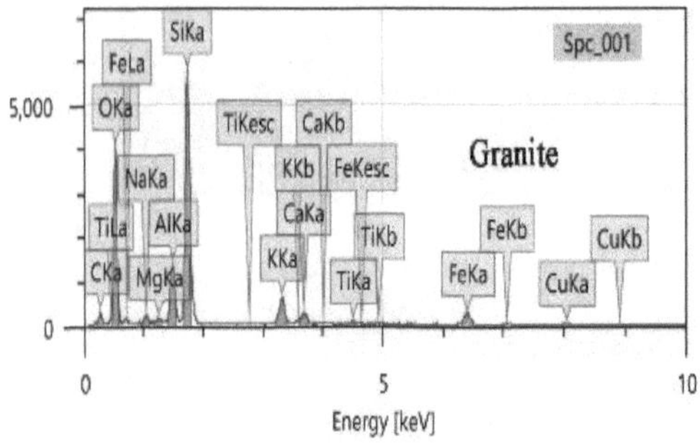

Display name	Standard data	Quantification method	Result Type
Spc_001	Standardless	ZAF	Metal

Element	Line	Mass%	Atom%
C	K	9.83±0.15	15.98±0.24
O	K	46.90±0.29	57.21±0.35
Na	K	1.33±0.05	1.13±0.04
Mg	K	0.50±0.03	0.40±0.02
Al	K	5.45±0.08	3.94±0.06
Si	K	22.85±0.15	15.88±0.11
K	K	4.14±0.07	2.07±0.04
Ca	K	1.94±0.05	0.94±0.03
Ti	K	0.60±0.03	0.24±0.01
Fe	K	5.03±0.11	1.76±0.04
Cu	K	1.42±0.08	0.44±0.03
Total		100.00	100.00
Spc_001			Fitting ratio 0.0145

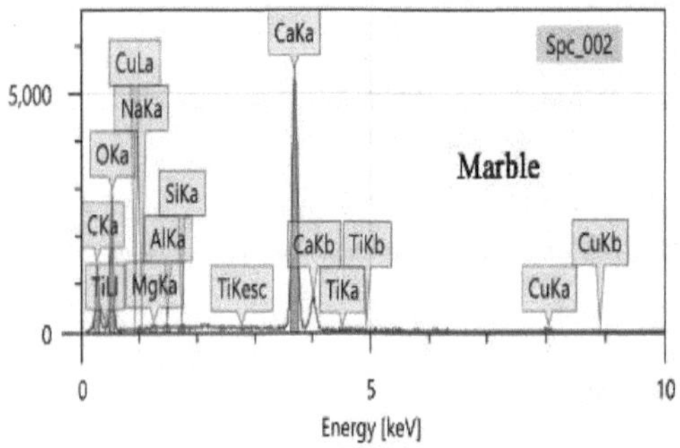

Display name	Standard data	Quantification method	Result Type
Spc_002	Standardless	ZAF	Metal

Element	Line	Mass%	Atom%
C	K	13.43±0.09	21.18±0.14
O	K	53.33±0.39	63.13±0.46
Na	K	0.03±0.02	0.02±0.02
Mg	K	0.26±0.02	0.21±0.02
Al	K	0.13±0.02	0.09±0.01
Si	K	0.06±0.02	0.04±0.01
Ca	K	31.86±0.17	15.06±0.08
Ti	K	0.04±0.02	0.02±0.01
Cu	K	0.84±0.06	0.25±0.02
Total		100.00	100.00
Spc_002			Fitting ratio 0.0183

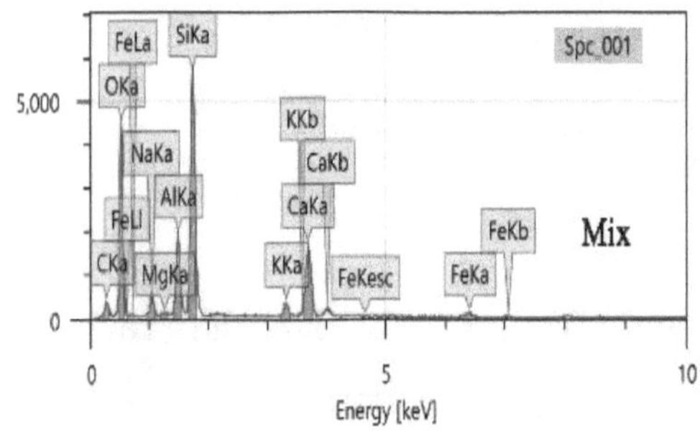

Display name	Standard data	Quantification method	Result Type
Spc_001	Standardless	ZAF	Metal

Element	Line	Mass%	Atom%
C	K	9.39±0.12	14.96±0.19
O	K	49.98±0.29	59.76±0.35
Na	K	2.46±0.06	2.04±0.05
Mg	K	0.23±0.02	0.18±0.02
Al	K	5.92±0.07	4.20±0.05
Si	K	18.59±0.13	12.66±0.09
K	K	1.65±0.04	0.81±0.02
Ca	K	10.04±0.10	4.79±0.05
Fe	K	1.75±0.06	0.60±0.02
Total		100.00	100.00
Spc_001			Fitting ratio 0.0165

Figure (5): EDX of granite, marble and granite: marble (2:1)

3.3. Study of the best adsorbents for the efficiency of lead and cadmium removal.

3.3.1. Adsorption of lead and cadmium

Table 4 shows the removal percentage of lead and cadmium in a single standard solution of ions. 50 ppm for Pb^{2+} and Cd^{2+} at temperatures 25, 50, and 60 °C.

Table (4): Removal efficiency of lead and cadmium ions by five prepared adsorbents (pH 7, 50ppm, 50 ml, contact time 1hr, and 0.5 g dose).

Adsorbents	Pb^{2+}					Cd^{2+}
	Removal %			Removal %		
Material	25 °C	50 °C	60 °C	25 °C	50 °C	60 °C
Granite	99.08	98.96	98.88	95.36	96.31	97.65
Marble	99.04	98.90	98.79	95.45	96.81	97.31
Granite: marble (1:1)	99.05	99.00	98.70	96.00	97.56	98.1
Granite: marble (1:2)	99.06	99.00	98.70	93.18	93.27	94.25
Granite: marble (2:1)	99.11	99.08	98.74	97.63	98.027	98.36

The results in Table 4, indicated that the removal percentage of lead changed with changing adsorbents and with alter temperature. The lower metal removal is 98.88% and the higher one is 99.11% with the mixture granite: marble (2:1) at 60 °C. The removal percentage of cadmium changes with changing adsorbents and with different temperatures; the lower metal removal is 95.36% and the higher one is 98.36% with a mixture of metal removal (2:1) at 60 °C.

Alwared et al.(2019) reported that under ideal circumstances, the maximum removal of 50 mg/l of Pb^{2+} ions in a single system using sorptive flotation with marble wastes was approximately 99.95% . **Mlayah et al.(2015)** investigated Bianco Gioia marble wastes (BGMW), a dose of 20 g/l was necessary to ensure the complete removal of lead.

Ghazy et al.(2008) reported that metal removal recovery of Cd^{2+} ions added to some water samples using 2000 mg/l of PMW sorbent at pH 7 (20 mg/l of Cd^{2+} in distilled water) was 99.45%. However, the lower values of some samples were enhanced to reach about 100% by increasing the PMW dose. **Alwared et al.(2019)** investigated that the maximum removal of 50 mg/l of Cd^{+2} ions in a single system using sorptive flotation with marble wastes was 78.697% at the best conditions. **Mlayah et al.(2018)** reported that the initial cadmium content of 5 mg l^{-1}, the adsorbent dosage of 15 g l^{-1}, and the flow rate of 20 ml min^{-1} were found to be the ideal parameters for the maximum cadmium removal rate by PMW (96.04%).

3.4. Studying the optimal conditions for adsorption of lead and cadmium

In these experiments, granite: marble (2:1) by (w/w) was chosen for the adsorption of Pb^{2+} and Cd^{2+} ions from water samples.

3.4.1. Initial lead and cadmium concentration on adsorption efficiency

The adsorption of granite: marble adsorbent was studied by adjusting the concentrations of Pb^{2+} and Cd^{2+} from 10 to 100 ppm at a constant temperature of 25 °C. Figure (6) shows that the initial concentration affected strongly the adsorbent's ability to remove Pb^{2+} and Cd^{2+}. The removal percentage increases from 96.6% to 99.11%, be constant up to

99% for Pb^{2+}. While for the removal of Cd^{2+} give a stable removal from 98.24% to 98.4%, then reached to 98.00% at high concentration. This is because the adsorbent's surface area and active sites get saturated at higher concentrations. As a result, the active adsorption sites that are required for adsorption weaken. Conversely, the initial concentration offers the necessary velocity needed to get beyond the mass transfer resistance between the solid adsorbent phase and the aqueous adsorbate phase. Because of this, there are more ions competing for accessible sites on the surface of the adsorbent when the initial pollutant load is large, which results in a higher adsorption capacity. **Tony(2020)** verified this occurrence in the process of treating textile wastewater by employing sludge as an adsorbent substance. The large ratio of metal ions to active sites at low metal concentrations promoted the heavy metal ions' adsorption, increasing the effectiveness of the removal. The diffusion velocity of heavy metal ions would decrease as metal ion concentrations increased **(Erdem et al., 2004).**

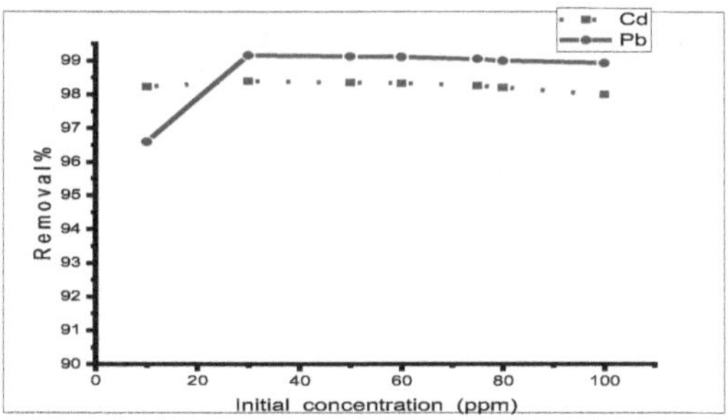

Figure (6): Initial concentration effect on removal of Pb^{2+} and Cd^{2+} using granite: marble (2:1).

3.4.2. pH effect on adsorption of lead and cadmium

The entire adsorption process is influenced by the pH of the Pb^{2+} and Cd^{2+} solutions. The effect of the pH of the solution on the adsorption of Pb^{2+} and Cd^{2+} on granite (2:1) was studied at metal concentrations (50 ppm), and temperature (25°C). Figure (7), displayed the data, indicated which indicated that the removal percentage of Pb^{2+}, and Cd^{2+} increase from 2 to 6. The adsorption removal percentage increased for Pb^{2+} from 47.24% to 99.05%, then decreased to 90% for lead while from Cd^{2+} ions increased from 24.84 %to 99.38% with increasing pH values. So, the best adsorption of Pb^{2+} and Cd^{2+} was pH 6. Research by **GM(1988) and Puranik et al.(1997)** shows that heavy metals removal became difficult at low pH values (less than 4.0), that might be explained by a high proton concentration in solution producing a charge that is positive density on binding metal sites. The metal binding's deprotonation sites cause an increase in the density of negative charges on the surface of carbon, which in turn promotes adsorption as pH rises. While at higher pH values than 8, metal ions may be precipitated, so all experiments were conducted with pH values of 7. The study by **Huang et al.(2017)** notes that higher pH values result in increased heavy metal adsorption. According to **Bedemo et al.(2016)**,higher pH levels result in a greater quantity of hydroxide ions (OH^-) and an increase in adsorption effectiveness because they dehydrate the functional groups that are present at the surface.

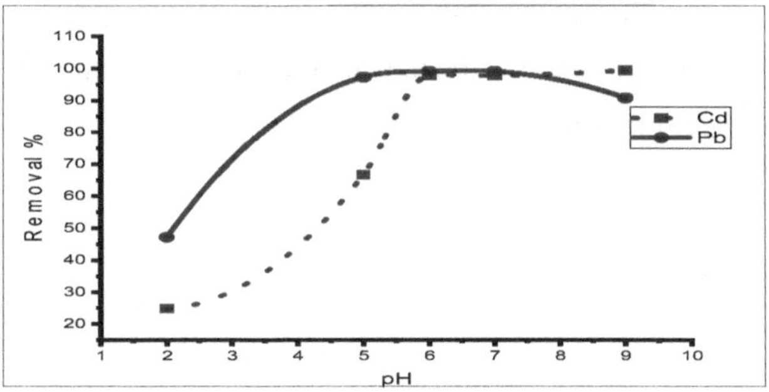

Figure (7): Effect of pH on removal of Pb^{2+} and Cd^{2+} using granite: marble (2:1).

3.4.3. Adsorbent dosage effect on adsorption of lead and cadmium

The effect of adsorbent dosage on Pb^{2+} and Cd^{2+} removal shows that as the dosage of granite: marble (2:1) adsorbent increases, so does the removal percentage of Pb^{2+} and Cd^{2+}. Lead from 60.0% to 99.48% with an increase in adsorbent dose to 0.25 gm. The percentage of Cd^{2+} removed increased from 30.0% to 99.8% with an increase in adsorbent dose to 1.0 g (Figure 8). The rise in the number of adsorbents readily available for adsorption sites led to the removal of heavy metals. Pb^{2+} removal percentages decreased when doses increased over 0.25 gm because buildup on the adsorbent continued for a while after that point. And up to 1.0 gm for cadmium. The suitable dose for Pb^{2+} removal was 0.1 gm, and for cadmium, 0.5 gm when using 50 ml of solution 50 ppm. Because there are more active sites on the adsorbent when the dosage of the adsorbent is increased, more surface area is available for adsorption; nevertheless, after equilibrium was established, increasing the number of active sites had no further effect **(Gerçel et al., 2007).**

Marín et al. (2009) said the findings may be explained by an increase in the surface areas of the adsorbent, which would increase the quantity of accessible adsorption sites

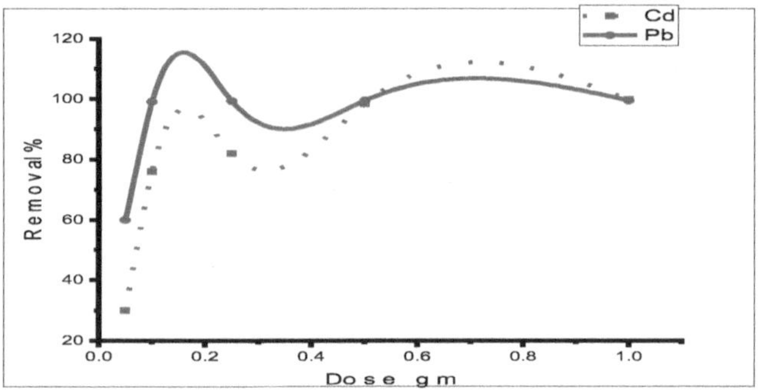

Figure (8): Adsorbent dosage effect on removal of Pb^{2+} and Cd^{2+} using granite: marble (2:1).

3.4.4. Time contact effect on adsorption of lead and cadmium

While maintaining the other parameters constant, the value of the Pb^{2+} and Cd^{2+} adsorption was altered during a series of experiments, ranging from 30 to 120 minutes. Figure (9) shows how contact duration affected the elimination of Pb^{2+} and Cd^{2+}. This suggests that there were two steps to the sorption of metal ions: a rapid sorption uptake that happened within 60 minutes of the sorbate-sorbent interaction. After that, there was a gradual phase of metal ion removal that lasted for 60 minutes until 120 minutes, at which point it was thought to have reached an equilibrium state. The optimal contact time for Pb^{2+} and Cd^{2+} for adsorption on granite and marble adsorbent was 60 minutes, according to the results. **Senberber et al.(2017)** used borax sludge to study the impact of adsorption contact time on Cr (III) adsorption from an aqueous solution. They discovered that q, or the amount of metal ions adsorbed, is very high during the initial stages of the Cr (III) adsorption processes and then decreases until it reaches equilibrium. The adsorption equilibrium was detected following the adsorption of hazardous cadmium, copper, and lead ions onto the adsorbent surface after four hours. **(Lee et al., 2018).**

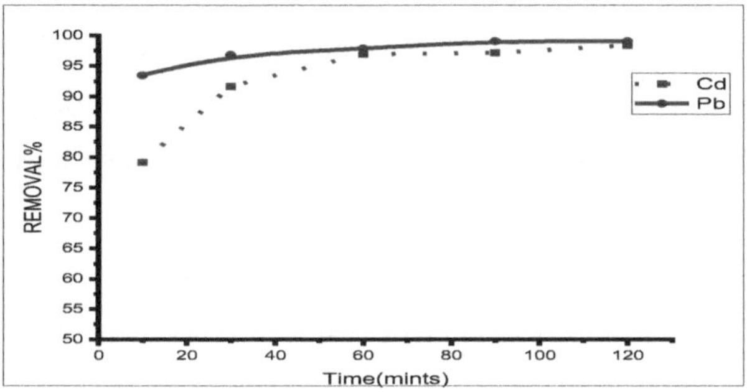

Figure (9): Time effect on removal of Pb^{2+} and Cd^{2+} using granite: marble (2:1).

3.4.5. Temperature effect on adsorption of lead and cadmium

At various temperatures, the influence of temperature on the adsorption of Pb^{2+} and Cd^{+2} has been investigated (25 °C, 40 °C, 50 °C, and 60 °C). The outcomes are shown in Figure (10). The findings indicated that the percentage of removals of Cd^{+2} on granite: marble adsorption grew as the temperature rose. For lead, the removal increased to 50 °C, then decreased. Because temperature has the potential to increase the rate of removal through activation energy, it is regarded as a crucial adsorption characteristic. When the temperature rises and increases the adsorption capacity or not, the adsorption system might be endothermic or exothermic, respectively **(Parker et al.,**

2013; Öztürk et al., 2014). The impact of increasing temperature on the adsorption of various dye molecules in chitin and the adsorption of Astrazone Blue dye on filler earth and fried clay adsorbents was examined by **Mckay et al.(1983)and McKay et al.(1985).** Ni(II) had a maximum equilibrium uptake of 78.9% at 293 K, after which it decreased linearly as temperature increased **(Gupta et al., 2019).** The high temperature makes it easier for ions to escape from the aqueous phase and lowers the boundary layer's mass transfer barrier. Usually, to learn more about the impact of additional contributing components on the adsorption process, the temperature at which maximum adsorption occurs is chosen. Furthermore, several researchers noted that neither a substantial rise nor fall in the absorption of heavy metals was associated with variations in temperature **(Bilal et al., 2021).**

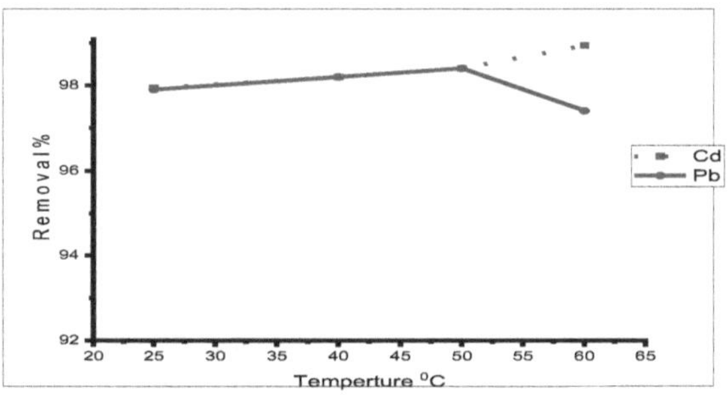

Figure (10): Temperature effect on removal of $Pb^{2+}$and Cd^{2+}using granite: marble (2:1).

3.5. Adsorption isotherms

Using equilibrium models and sorption isotherms, one can investigate the interactions between an adsorbent and an adsorbate **(Ai et al., 2011).** Langmuir, Freundlich, Dubinin-Raduskevich (D-R), linear, and

Temkin models are common equilibrium models used to describe isotherms of metal sorption on granite and marble.

35.1. **Linear isotherm model**

Henry's isotherm model is thought to be the most basic adsorption isotherm model since the amount of surface adsorbate is directly proportional to the partial pressure of the adsorptive gas, according to **Ayawei et al.(2017).** Henry's adsorption isotherm model, in which every adsorbate molecule is isolated from its nearest neighbor, describes an acceptable fit of the adsorbate adsorption at low concentrations. This model applies to liquid solutions.

The linear connection presents the linear model, which characterizes the solute accumulation by the sorbent as being exactly proportional to the solution concentration.

$$qe = Kd.\ Ce \quad (2)$$

Where C_e is the metal's equilibrium concentration, K_d is the equilibrium distribution coefficient, and q_e is the amount of adsorbate (mg/g). A straight line was generated by graphing q_e against C_e, and K_d was calculated from the slope.

3.5.2. Langmuir isotherm model

Probably the most well-known and commonly employed sorption isotherm model is the Langmuir model. Among the core tenets of this mechanistic model is monolayer adsorption, which maintains that sorption is restricted to a finite number of identical active sites (the adsorbed layer has a thickness of one molecule) **(Gerente et al., 2007).**

The following equation characterizes the Langmuir equation's linearized form:

$$Ce/qe=1/Qobl+Ce/Qo \quad (3)$$

b_l is a Langmuir binding energy constant (l/mg), q_e is the heavy metals' equilibrium capacity (mg/g) on the adsorbent, and Q_o is the adsorbent's monolayer adsorption capacity (mg/g). C_e is the equilibrium concentration of the heavy metals solution (mg/l), Q_o is the adsorbent's capacity for monolayer adsorption (mg/g) and b is the Langmuir binding energy coefficient (l/mg) **(Chakraborty et al., 2022).** The linear plot of C_e/q_e against C_e can be used to compute the b and Q_o based on its intercept and slope.

3.5.3. Freundlich isotherm

The Freundlich isotherm is another well-known isotherm that is widely used to depict adsorption equilibrium at a constant temperature. The Freundlich equation determines how heterogeneous surface energy is used in the adsorption process.

The empirical Freundlich equation has the following expression:

$$\text{Log } q_e = \log K_f + 1/n \log C_e \quad (4)$$

Where K_f ($mg^{-1/n}$ $L^{1/n}/g^{-1}$) is the adsorption capacity indicated by the Freundlich isotherm constant and n is the Freundlich equilibrium constant that expresses the binding energy between the adsorbent and metal ions. Normal adsorption is indicated by 1/n values less than one, whereas cooperative adsorption is indicated by 1/n values more than one. The linear plot of log q_e against log C_e can be used to compute K_f and n based on its intercept and slope.

3.5.4. Temkin Isotherm Model

A homogeneous distribution of binding energy over the population of surface binding adsorption is predicted by the Temkin isotherm model. Additionally, it illustrates a monolayer distribution of energy throughout the adsorption surface's active places. This isotherm implies that the adsorption in the layer decreases with coverage due to interactions between the adsorbent material and the adsorbate. The Temkin equation is represented in its linear form by

$$q_e = B \ln K + B \ln C_e \quad (5)$$

In this equation, B is equal to RT/b, where b is the Temkin constant and q_e (mg/g) is the heat of sorption is the amount of metal ions adsorbed per unit weight of the adsorbent. The metal ion concentration in solution at equilibrium is expressed as C_e (ppm), the equilibrium binding constant of the Temkin isotherm is expressed as K, and the heat of sorption is expressed as B (j/mol). Furthermore, one can ascertain the values of B and K by plotting q_e versus ln C_e.

4.2.2.5. Dubin-Raduskevich Isotherm Model

The Dubinin-Radushkevich (D-R) model operates under the assumption that adsorption is dependent on both pore volume and surface porosity. To examine the adsorption via an energy-based view. The equation of the Dubinin-Radushkevich (D-R) model is given below.

$$\text{Ln } q_e = \ln q_m - BE^2 \quad (6)$$

where, B is a constant associated with the adsorption energy (mol/kJ), and q_m is a value that represents the sorption degree characterizing the sorbent (mg/g), E is the Polanyi potential, which can be found by solving the following equation:

$$E = RT \ln (1+1/C_e) \ (7)$$

Where R is the ideal gas constant ($R = 8.314$ j mol^{-1} K^{-1}), and T is the absolute temperature (K). Where q_e is the equilibrium capacity of the heavy metals on the adsorbent (mg/g) and C_e is the equilibrium concentration of the heavy metal solution (mg/l), Where B is denoted as the isotherm constant E can be computed by the relationship.

$E=1/ \quad (2B)^{½}$ (8)

If E is less than 8 KJ/mol, the sorption process happens physically; if E is between 8 and 16 KJ/mol, the sorption process occurs chemically **(Kiran et al., 2006; Sawalha et al., 2006)**.

For the adsorption of Pb^{2+} and Cd^{2+}, Table 5 includes all the isotherm model parameters. The estimated adsorption max capacities for lead and cadmium derived from the Langmuir isotherm were 171.23 mg/g and 36.37 mg/g, respectively. The computed values of 1/n for lead and cadmium derived from Freundlich isotherm, were 0.82623 and 0.85118, respectively, which show the degree of favorability of adsorption as well as effective adsorption. Lower values of K (cadmium and lead) and higher values of B (Binding of isotherm equilibrium) indicate that the Temkin model is followed by the adsorption of the adsorption energies of lead and cadmium are determined to be 2.37 and 1.74 kJ/mol using the D-R isotherm. These low adsorption energy values demonstrate that low adsorption energy values demonstrated the physical nature of the adsorption process. R^2 were for the four models (linear, Langmuir, Freundlich, Dubinin-Raduskevich (D-R), and Temkin models) 0.9995,0.9834,0.9704, and 0.9606, respectively, for lead, while for cadmium 0.9775,0.9230,0.9911, and 0.9848, respectively.

Table (5): Adsorption isotherms calculated parameters for adsorption of Pb^{2+}, Cd^{2+} ions on granite: marble (2:1).

parameter	Pb^{2+}	Cd^{2+}
Linear Modal		
K_d (l/ g)	41.99	4.50
R^2	0.9995	0.9775
Model of Langmuir isotherms		
Q_o (mg/g)	171.23	36.37
b_L(l/mg)	0.3765	0.1901
R_L	0.1527	0.1263
R^2	0.9834	0.9230
Model of Freundich isotherm		
1/n	0.8262	0.8511
K_f $(mg^{-1/n}\ l^{1/n}/g^{-1})$	47.356	5.7236
R^2	0.9987	0.9911
Model of Temkin isotherm		
B(j/mol)	23.784	4.818
b_T(l/g)	104.16	514.20
K_t (l/g)	6.755	3.591
R^2	0.9704	0.9848
Model of Dubin-Raduskevich Isotherm		
q_m(mg/g)	56.4243	10.3630
E(kj/mol)	2.37	1.74
B(mol/kj^2)	8.90E-08	1.65E-07
R^2	0.9606	0.9563

3.6. Adsorption kinetic studies

The adsorption process's rate-limiting phases can be studied by utilizing the contact time found in experimental data. The rate can be determined by kinetic analysis, and these are the most crucial phases. The most widely utilized kinetic models to describe adsorption kinetic studies are Lagergen's first-order equation and Ho's second-order equation. The kinetic models that follow illustrate how metal ion kinetics were analyzed on various adsorbent materials **(Chakraborty et al., 2022)**.

Processing adsorption kinetics data is the best way to comprehend the dynamics of the adsorption process. It aids in the estimation of the adsorption rate, providing data for process modeling and design **(El Nemr, 2009)** . Two main adsorption processes could occur throughout the adsorption system: physisorption, which is weak as it depends on van der Waals attraction forces, and chemisorption, due to the strong link created between the solute and the adsorbent that involves electron transfer. The information gathered from figuring out the reaction rates might help in the creation of adsorbent materials for use in industry. Additionally, it might be beneficial for comprehending the intricate dynamics of the sorption process well **(Tan et al., 2017).**

3.6. 1. Pseudo–first- order model

Based on the solid capacity**, Lagergren(1898)** proposed a first-order reaction equation for the sorption of liquid and solid systems. This formula is frequently used to determine adsorption rates in liquid-phase investigations, especially when it comes to the adsorption of dyes and the removal of metal ions from wastewater. This model suggests that metal adsorption is first order in nature since it is predicated on the amount of metal ions in the solution at any time. The linear pseudo-first-order equation is in the form.

$$\text{Log } (q_e - q_t) = \log q_e - (K_1/2.303)\, t \quad (9)$$

Where the amounts of metal ions adsorbed onto the biosorbent at equilibrium and time (min) are denoted by q_e and q_t (mg/g), respectively. K1 is the pseudo-first-order rate constant. Plotting log (q_e-q_t) vs. time (min) yielded a linear relationship with slope k1 and intercept log q_e, signifying the pseudo-first-order rate constant K_1.

4.2.3.2. Pseudo–second order model

Experiments are required to determine the rate constant and reaction order. Using a pseudo-second-order formula for the rate law, it was possible to see that the rate was dependent upon the sorption capacity rather than the sorbate concentration **(Ho et al., 2000).** The removal from a solution in this model is caused by physicochemical interactions between the two phases during surface adsorption, which is the rate-limiting step **(Wang et al., 2007).** This is the linearized form of the pseudo-second-order kinetic model:

$$t/q_t = 1/(K_2 q_e^2) + 1/q_e\, t \quad (10)$$

Where K_2 is the pseudo-second-order adsorption rate constant (g/mg min). Plotting t/q_t vs. t reveals a linear relationship; the plot's intercept and slope are used to derive the values of K_2 and equilibrium adsorption capacity, or q_e.

4.2.3.3. Intra-particle diffusion model

The solute transfer in a solid-liquid adsorption process is commonly depicted using the intra-particle diffusion model, which is employed to ascertain the adsorption process's mechanism. **(Weber and Morris,**

1963). There is multilinearity in the plot of q_t vs. $t^{1/2}$; each component corresponds to a different mass transfer process. The immediate adsorption stage, also known as film diffusion or external mass transfer, is the first part. The slow adsorption phase, which makes up the second section, is when intra-particle diffusion may have rate-limiting effects. The third section is the last equilibrium stage, where the extremely low solute concentration in the solution causes the intra-particle diffusion to slow down. **Noroozi et al. (2007)** used the intra-particle diffusion model and the following equation:

$$q_t = K_{ip}\, t^{1/2} + C \qquad (11)$$

Where C (mg g^{-1}) and the intra-particle diffusion model's rate constant is K_{ip} (mg g^{-1} $min^{-1/2}$).

represents the impact of the boundary layer or the thickness of the boundary layer. The correlation coefficient (R^2), boundary layer effect (C), and intra-particle diffusion constant (k_{ip}) are computed from a plot of q_t vs. $t^{1/2}$.

Table (6) shows that the calculated equilibrium adsorption capacity (q_e) derived from pseudo second order model curve for Pb^{2+} and Cd^{+2} ions on granite and marble mixture were 24.876 and 4.994 mg/g, respectively, which nearly match the experimental q_e results. The calculated K_{ip} (g/mg/min) values for Pb^{2+} and Cd^{+2} ions are 0.060 and 0.085, respectively. Furthermore, the correlation coefficient (R^2) for the pseudo-second order than the first order and equal 1.00. Therefore, the adsorption of Pb^{2+} and Cd^{+2} ions on granite and marble adsorbents fits the- pseudo -second-order kinetic. The data showed that the C (the boundary layer effect) of Pb^{2+} and Cd^{+2} is 24.371 and 4.630 mg g^{-1}, respectively. The first line would depict the quick absorption and physical forces that cause the adsorbate to transfer quickly onto the adsorbent surface. Since the line does not cross the origin in the first stage, it is notable that film diffusion dominates uptake rather than the intra-particle diffusion process. While the third stage displays the saturation of the adsorbent surface, the second stage of sorbate adsorption accelerates, reflecting non-consecutive diffusion of sorbate molecules into the micropores with pore width within the sorbent. The high levels of the regression coefficients also suggested that particle diffusion might play a role in the sorption of heavy metals. Typically, the preferential adsorption of sorbate in the micropores is the cause of the adsorption regulated by the intraparticle model **(Biyan et al., 2009)**. The intercept value from these data shows that the lines are not going through the origin, indicating the possibility of another process affecting the adsorption.

Table (6): Kinetic parameters for the adsorption were calculated of Pb^{2+}, and Cd^{+2} ions on granite: marble (2:1).

parameter	Pb^{2+}	Cd^{2+}
Pseudo- first- order model		
k_1(min) $^{-1}$	0.0299	0.053
q_e(calculated)mg/g	1.1732	1.623
R^2	0.9685	1.000
Pseudo- second -order model		
k_2 (g/mg min)	0.060	0.085
q_e(calculated)mg/g	24.876	4.994
q_e(expemintal)mg/g	24.778	4.918
R^2	1.000	1.000
Intra- particle diffusion model		
k_{ip}	0.0484	0.0038
c(mg/g)	24.161	4.450
R^2	0.8594	0.73

4.2.4. Thermodynamics of Adsorption of lead and cadmium

Thermodynamic models are effective instruments for characterizing metal sorption procedures and investigating regulating mechanisms. The adsorption of metal ions related to the thermodynamics of the adsorption process is significantly influenced by temperature. Exothermic or endothermic adsorption processes can be used to describe the thermodynamic nature of heavy metal adsorption on inexpensive adsorbents **(Inyang et al., 2016).** An endothermic process occurs when adsorption rises with rising temperatures, whereas an exothermic process occurs when adsorption decreases with rising temperatures. An endothermic process occurs when adsorption increases with increasing temperatures, whereas an exothermic process occurs when adsorption decreases with rising temperatures. Adsorption thermodynamic behavior can be determined using three thermodynamic parameters, namely enthalpy (ΔH°), Gibbs free energy (ΔG°), and entropy (ΔS°) **(Yeganeh et al., 2019).** The following equations are used:

$$k_d = q_e / c_e \qquad (11)$$

It is possible to express the equilibrium constant K in terms of the concentrations of each medium. Thermodynamic properties like changes in enthalpy (ΔH°), entropy (ΔS°), and Gibbs free energy (ΔG°) are the true indicators for particle applications. Four different temperatures were used to evaluate the thermodynamics of adsorption (298, 313, 323, and 333 K). The following equation was used to compute the thermodynamic parameters.

$$\text{Ln } k_d = (\Delta S^o / R) - (\Delta H^o / RT) \qquad (12)$$

T is the absolute temperature in Kelvin (K_d), and the universal gas constant is R (8.314 J/mol K). The van't Hoff plots' intercept and slope of ln k versus 1/T were used to calculate both (ΔH°) and (ΔS°) (**Nollet et al., 2003; Özcan et al., 2006).** ΔG^0 (KJ/mol), the free energy of selective adsorption, can be computed using the subsequent expression:

$$\Delta G^0 = - RT \ln K_d \qquad (13)$$

From Table 7, the endothermic nature of the adsorption process is shown by a positive enthalpy (ΔH°) value (Cd^{2+}), whereas the exothermic nature of the process is indicated by a negative (ΔH°) value. (Pb^{2+}) Conversely, a negative value of (ΔG°) suggests that the adsorption process is thermodynamically possible and spontaneous at all temperatures. During the adsorption process, a positive value of (ΔS°) indicated greater unpredictability at the solid-liquid interface, whereas a negative value of (ΔS°) suggests decreased randomness. The low value of ΔH° suggested that this adsorption system may operate mainly by physical adsorption. The ΔS° was determined to be between 0.0116 and 0.096 J/molK^{-1}. When ion molecules adsorb on the active sites of the adsorbent surface, there is a rise in randomness at the solid-liquid interface, which leads to a positive value of ΔS°.

Tables (7) Thermodynamic data of the adsorption of Pb^{2+}, and Cd^{+2} on granite: marble (2:1).

parameters		Pb^{2+}	Cd^{2+}
ΔH°, K J/mol		-4.44	23.64
ΔS°, K J/mol K		0.0116	0.0967
ΔG°K J/mol	298K	-7.799	-5.160
	313K	-8.219	-6.797
	323K	-8.404	-7.336
	333K	-8.112	-8.721
R^2		0.7612	0.9792

5. **Application in wastewater (real sample test)**

Under ideal sorption conditions that were adjusted at constant values (adsorbent dose 0.1g, contact time 2h at 250 rpm, pH6 at 25°C, and 50 ml of solution), the efficacy of this adsorbent examined in removing Pb^{2+}, and Cd^{2+} uptake. The intention of putting a mixture of granite and marble had been used in the wastewater contained of 0.441 and 0.25

ppm of Pb^{2+}, and Cd^{2+} respectively. These variables were altered following the addition of the adsorbent to the mixture.

The wastewater sample came from Kima's factory disposal lab. Several parameters were measured before the adsorption process began, including temperature, electrical conductivity, pH, total dissolved solids, and the concentrations of Pb^{2+}, and Cd^{2+}ions. Following the adsorption procedure, the concentrations of the ions Pb^{2+}, and Cd^{2+} were determined. After calculating the percentage of ions removed from the solution, the findings were recorded. The proportion of Pb^{2+}, and Cd^{2+} ions removed from the granite: marble adsorbents approached 99.48, and 99.8%,

respectively. These findings validate the granite: marble's great efficacy in eliminating heavy metals from laboratory waste, as indicated in Table (8).

Table (8): Adsorption removal percentage for Pb^{2+}, and Cd^{2+} of the real wastewater on granite: marble (2:1)

ITEM	Pb^{2+}	Cd^{2+}
Concentration before adsorption (ppm)	0.441	0.25
Concentration after adsorption (ppm)	0.00229	0.0005
Removal%	99.48	99.8

5. Conclusion

In this research, an efficient adsorbent was made by mixing granite: marble (2:1), which, exhibited excellent efficiency in removing Pb^{2+} and Cd^{2+} ions from water up to 99%. The adsorption process is dependent on pH, contact time, adsorbent dose, initial metal concentration, and solution temperature. According to the findings of experimental data for adsorption of Pb^{2+} and Cd^{+2}, the kinetics were followed well the pseudo-second-order kinetic model. The Langmuir model fitted well with tested metal adsorption. The adsorption of lead and cadmium ions by a mixture of granite: marble (2:1) adsorbents were shown a negative value of ΔG^{o} for both Pb^{2+} and Cd^{+2} removal. This adsorbent's is considered eco-friendly; easy-to-use process would an important method for treating industrial wastewater.

References

Ai, L., Zhang, C., Liao, F., Wang, Y., Li, M., Meng, L., and Jiang, J. (2011). Removal of methylene blue from aqueous solution with magnetite loaded multi-wall carbon nanotube: kinetic, isotherm and mechanism analysis. *Journal of hazardous materials, 198*, 282-290.

Alwared, A. I., and Sadiq, N. A. (2019). Competitive removal of lead copper and cadmium ions by sorptive flotation using marble wastes. *International Journal of Environment and Waste Management, 23*(2), 156-178.

Ayawei, N., Ebelegi, A. N., and Wankasi, D. (2017). Modelling and interpretation of adsorption isotherms. *Journal of chemistry, 2017.*

Batheja, K., Sinha, A., and Seth, G. (2009). Studies on water treatment for removal of nitrate. *Asian J. Exp. Sci, 23*(1), 61-66.

Bedemo, A., Chandravanshi, B. S., and Zewge, F. (2016). Removal of trivalent chromium from aqueous solution using aluminum oxide hydroxide. *SpringerPlus, 5*, 1-11.

Bilal, M., Ihsanullah, I., Younas, M., and Shah, M. U. H. (2021). Recent advances in applications of low-cost adsorbents for the removal of heavy metals from water: A critical review. *Separation and purification technology, 278*, 119510.

Biyan, J., Fei, S., Hu, G., Zheng, S., Zhang, Q., and Xu, Z. (2009). Adsorption of methyl tert-butyl ether (MTBE) from aqueous solution by porous polymeric adsorbent. *J. Hazard. mater, 161*(1), 81-87.

Brunauer, S., Emmett, P. H., and Teller, E. (1938). Adsorption of gases in multimolecular layers. *Journal of the American chemical society, 60*(2), 309-319.

Chakraborty, R., Asthana, A., Singh, A. K., Jain, B., and Susan, A. B. H. (2022). Adsorption of heavy metal ions by various low-cost adsorbents: a review. *International Journal of Environmental Analytical Chemistry, 102*(2), 342-379.

Chronopoulos, J., Haidouti, C., Chronopoulou-Sereli, A., and Massas, I. (1997). Variations in plant and soil lead and cadmium content in urban parks in Athens, Greece. *Science of the total environment, 196*(1), 91-98.

Diagomanolin, V., Farhang, M., Ghazi-Khansari, M., and Jafarzadeh, N. (2004). Heavy metals (Ni, Cr, Cu) in the karoon waterway river, Iran. *Toxicology letters, 151*(1), 63-67.

Diouri, K., Chaqroune, A., Kherbeche, A., Miyah, Y., and Lahrichi, A. (2014). Kinetics of Direct Yellow 50 dye adsorption onto marble powder sorbents. *International Journal of Innovative Research in Science Engineering and Technology, 3*, 16626.

Dureja, H.,and Madan, A. (2007). Adsorption in pharmacy: A review. *Drug Delivery Technology, 7*(8), 58-64.

El Nemr, A. (2009). Potential of pomegranate husk carbon for Cr (VI) removal from wastewater: Kinetic and isotherm studies. *Journal of hazardous materials, 161*(1), 132-141.

Erdem, E., Karapinar, N., and Donat, R. (2004). The removal of heavy metal cations by natural zeolites. *Journal of colloid and interface science, 280*(2), 309-314.

Familusi, A. O., ADEKUNLE, A., Badejo, A., and Adeosun, O. J. (2021). Distinctive Features of River Sand, Beach Sand and Granite as Adsorbents in Water and Wastewater Treatment. *The Egyptian International Journal of Engineering Sciences and Technology, 34*(1), 11-15.

Favier, A., Habert, G., De Lacaillerie, J. D. E., and Roussel, N. (2013). Mechanical properties and compositional heterogeneities of fresh geopolymer pastes. *Cement and Concrete Research, 48*, 9-16.

Gao, X., Yuan, B., Yu, Q., and Brouwers, H. (2017). Characterization and application of municipal solid waste incineration (MSWI) bottom ash and waste granite powder in alkali activated slag. *Journal of Cleaner Production, 164*, 410-419.

Gerçel, Ö., and Gerçel, H. F. (2007). Adsorption of lead (II) ions from aqueous solutions by activated carbon prepared from biomass plant material of Euphorbia rigida. *Chemical engineering journal, 132*(1-3), 289-297.

Gerente, C., Lee, V., Cloirec, P. L., and McKay, G. (2007). Application of chitosan for the removal of metals from wastewaters by adsorption—mechanisms and models review. *Critical reviews in environmental science and technology, 37*(1), 41-127.

Ghazy, S. (1995). Removal of cadmium, lead, mercury, tin, antimony, and arsenic from drinking and seawaters by colloid precipitate flotation. *Separation Science and Technology, 30*(6), 933-947.

Ghazy, S. E., Gabr, I. M., and Gad, A. H. (2008a). Cadmium (II) sorption from water samples by powdered marble wastes. *Chemical Speciation & Bioavailability, 20*(4), 249-260.

Ghazy, S. E., and Gad, A. (2008b). Separation of Zn (II) by sorption onto powdered marble wastes.

GM, G. (1988). Accumulation of metals by microorganisms and algae. *Biotechnology: a comprehensive treatise, 6*, 401-433.

Gunasekaran, S., and Anbalagan, G. (2007). Spectroscopic characterization of natural calcite minerals. *Spectrochimica Acta Part A: Molecular and Biomolecular Spectroscopy, 68*(3), 656-664.

Gupta, S., Sharma, S., and Kumar, A. (2019). Biosorption of Ni (II) ions from aqueous solution using modified Aloe barbadensis Miller leaf powder. *Water Science and Engineering, 12*(1), 27-36.

Hahladakis, J., Smaragdaki, E., Vasilaki, G., and Gidarakos, E. (2013). Use of sediment quality guidelines and pollution indicators for the assessment of heavy metal and PAH contamination in Greek surficial sea and lake sediments. *Environmental monitoring and assessment, 185*, 2843-2853.

Handke, M., and Mozgawa, W. (1993). Vibrational spectroscopy of the amorphous silicates. *Vibrational Spectroscopy, 5*(1), 75-84.

Ho, Y. S., and McKay, G. (2000). The kinetics of sorption of divalent metal ions onto sphagnum moss peat. *Water research, 34*(3), 735-742.

Huang, J., Yuan, F., Zeng, G., Li, X., Gu, Y., Shi, L., . . . Shi, Y. (2017). Influence of pH on heavy metal speciation and removal from wastewater using micellar-enhanced ultrafiltration. *Chemosphere, 173*, 199-206.

Hussain, A., Nallu, M., and Arivoli, S. (2012). Kinetics and thermodynamic studies of Copper (II) ion from aqueous solution by activated granite. *Int J Adv Scientific Res Technol, 3*(2), 69-78.

Inyang, M. I., Gao, B., Yao, Y., Xue, Y., Zimmerman, A., Mosa, A., and Cao, X. (2016). A review of biochar as a low-cost adsorbent for aqueous heavy metal removal. *Critical reviews in environmental science and technology, 46*(4), 406-433.

Kiran, I., Akar, T., Ozcan, A. S., Ozcan, A., and Tunali, S. (2006). Biosorption kinetics and isotherm studies of Acid Red 57 by dried Cephalosporium aphidicola cells from aqueous solutions. *Biochemical engineering journal, 31*(3), 197-203.

Kobya, M., Demirbas, E., Senturk, E., and Ince, M. (2005). Adsorption of heavy metal ions from aqueous solutions by activated carbon prepared from apricot stone. *Bioresource technology, 96*(13), 1518-1521.

Lagergren, S. K. (1898). About the theory of so-called adsorption of soluble substances. *Sven. Vetenskapsakad. Handingarl, 24*, 1-39.

Lane, T. W., and Morel, F. M. (2000). A biological function for cadmium in marine diatoms. *Proceedings of the National Academy of Sciences, 97*(9), 4627-4631.

Lee, S.-Y., and Choi, H.-J. (2018). Persimmon leaf bio-waste for adsorptive removal of heavy metals from aqueous solution. *Journal of environmental management, 209*, 382-392.

Marín, A. P., Aguilar, M., Meseguer, V., Ortuno, J., Sáez, J., and Lloréns, M. (2009). Biosorption of chromium (III) by orange (Citrus cinensis) waste: Batch and continuous studies. *Chemical engineering journal, 155*(1-2), 199-206.

Mckay, G., Blair, H., and Gardner, J. (1983). The adsorption of dyes in chitin. III. Intraparticle diffusion processes. *Journal of Applied Polymer Science, 28*(5), 1767-1778.

McKay, G., Otterburn, M. S., and Aga, J. A. (1985). Fuller's earth and fired clay as adsorbents for dyestuffs: equilibrium and rate studies. *Water, Air, and Soil Pollution, 24*, 307-322.

Merzouk, B., Yakoubi, M., Zongo, I., Leclerc, J.-P., Paternotte, G., Pontvianne, S., and Lapicque, F. (2011). Effect of modification of textile wastewater composition on electrocoagulation efficiency. *Desalination, 275*(1-3), 181-186.

Mizuta, K., Matsumoto, T., Hatate, Y., Nishihara, K., and Nakanishi, T. (2004). Removal of nitrate-nitrogen from drinking water using bamboo powder charcoal. *Bioresource technology, 95*(3), 255-257.

Mlayah, A., and Jellali, S. (2015). Study of continuous lead removal from aqueous solutions by marble wastes: efficiencies and mechanisms. *International Journal of Environmental Science and Technology, 12*, 2965-2978.

Mlayah, A., and Jellali, S. (2018). Powdered marble wastes reuse for cadmium removal from aqueous solutions under dynamic conditions. *Environmental Engineering & Management Journal (EEMJ), 17*(5).

Nibou, D., Mekatel, H., Amokrane, S., Barkat, M., and Trari, M. (2010). Adsorption of Zn2+ ions onto NaA and NaX zeolites: Kinetic, equilibrium and thermodynamic studies. *Journal of hazardous materials, 173(*1-3), 637-646.

Nollet, H., Roels, M., Lutgen, P., Van der Meeren, P., and Verstraete, W. (2003). Removal of PCBs from wastewater using fly ash. *Chemosphere, 53*(6), 655-665.

Noroozi, B., Sorial, G., Bahrami, H., and Arami, M. (2007). Equilibrium and kinetic adsorption study of a cationic dye by a natural adsorbent—Silkworm pupa. *Journal of hazardous materials, 139*(1), 167-174.

Özcan, A., Öncü, E. M., and Özcan, A. S. (2006). Kinetics, isotherm and thermodynamic studies of adsorption of Acid Blue 193 from aqueous solutions onto natural sepiolite. *Colloids and surfaces A: Physicochemical and engineering aspects, 277*(1-3), 90-97.

Öztürk, A., and Malkoc, E. (2014). Adsorptive potential of cationic Basic Yellow 2 (BY2) dye onto natural untreated clay (NUC) from aqueous phase: mass transfer analysis, kinetic and equilibrium profile. *Applied Surface Science, 299*, 105-115.

Pal, M., Ayele, Y., Hadush, M., Panigrahi, S., and Jadhav, V. J. (2018). Public health hazards due to unsafe drinking water. *Air Water Borne Dis*, *7*(1000138), 2.

Parker, H. L., Budarin, V. L., Clark, J. H., and Hunt, A. J. (2013). Use of starbon for the adsorption and desorption of phenols. *ACS Sustainable Chemistry & Engineering, 1*(10), 1311-1318.

Puranik, P., and Paknikar, K. (1997). Biosorption of lead and zinc from solutions using Streptoverticillium cinnamoneum waste biomass. *Journal of biotechnology, 55*(2), 113-124.

Sawalha, M. F., Peralta-Videa, J. R., Romero-González, J., and Gardea-Torresdey, J. L. (2006). Biosorption of Cd (II), Cr (III), and Cr (VI) by saltbush (Atriplex canescens) biomass: thermodynamic and isotherm studies. *Journal of colloid and interface science, 300*(1), 100-104.

Senberber, F., Yildirim, M., Mermer, N., and Derun, E. (2017). Adsorption of Cr (III) from aqueous solution using borax sludge. *Acta Chimica Slovenica, 64*(3).

Soliman, N. K. and A. Moustafa (2020). "Industrial solid waste for heavy metals adsorption features and challenges; a review." J*ournal of Materials Research and Technology 9(*5): 10235-10253.

Sitarz, M., Handke, M., and Mozgawa, W. (2000). Identification of silicooxygen rings in SiO2 based on IR spectra. *Spectrochimica Acta Part A: Molecular and Biomolecular Spectroscopy, 56*(9), 1819-1823.

Tan, K., and Hameed, B. (2017). Insight into the adsorption kinetics models for the removal of contaminants from aqueous solutions.

Journal of the Taiwan Institute of Chemical Engineers, 74, 25-48.

Tony, M. (2020). Zeolite-based adsorbent from alum sludge residue for textile wastewater treatment. *International Journal of Environmental Science and Technology, 17*(5), 2485-2498.

Tozsin, G. (2016). Inhibition of acid mine drainage and immobilization of heavy metals from copper flotation tailings using a marble cutting waste. *International Journal of Minerals, Metallurgy, and Materials, 23*, 1-6.

Wang, H., Zhou, A., Peng, F., Yu, H., & Yang, J. (2007). Mechanism study on adsorption of acidified multiwalled carbon nanotubes to Pb (II). *Journal of colloid and interface science, 316*(2), 277-283.

Wazwaz, A., Al-Salaymeh, A., and Khan, M. S. (2019). Removing heavy metals through different types of soils and marble powder found in Oman. *Journal of Ecological Engineering, 20*(4), 136-142.

Weber Jr, W. J., and Morris, J. C. (1963). Kinetics of adsorption on carbon from solution. *Journal of the sanitary engineering division, 89*(2), 31-59.

World Health Organization. (2000). Fifty-third report of the joint FAO/WHO expert committee on food additives. *WHO Technical Report Series, 896*, 1-136.

Xin, K. L. (2019). Removal of iron in groundwater using marble column filter. *International Journal of Integrated Engineering, 11*(2).

Yarrakula, K. (2019). Geo-chemical exploration of granite mining waste using XRD, SEM/EDX and AAS analysis. Iranian Journal of Chemistry and Chemical Engineering (IJCCE), 38(2), 215-228.

Yeganeh, G., Ramavandi, B., Esmaeili, H., and Tamjidi, S. (2019). Dataset of the aqueous solution and petrochemical wastewater treatment containing ammonia using low cost and efficient bio-adsorbents. *Data in brief, 26*, 104308.

Yılmaz, B., and Olgun, A. (2008). Studies on cement and mortar containing low-calcium fly ash, limestone, and dolomitic limestone. *Cement and Concrete Composites, 30*(3), 194-201.

Zacaroni, L. M., Magriotis, Z. M., das Graças Cardoso, M., Santiago, W. D., Mendonça, J. G., Vieira, S. S., and Nelson, D. L. (2015). Natural clay and commercial activated charcoal: Properties and application for the removal of copper from cachaça. *Food control, 47*, 536-544.

Table of content

FSC
www.fsc.org
MIX
Papier | Fördert gute Waldnutzung
FSC® C083411

Kolibri 360 GmbH
Ferdinand-Jühlke-Straße 7
99095 Erfurt, Deutschland
produktsicherheit@kolibri360.de

Druck:
CPI Druckdienstleistungen GmbH
im Auftrag der
Kolibri 360 GmbH
Ferdinand-Jühlke-Str. 7
99095 Erfurt